BRIAN GARDNER

The Difference Between Book Knowledge And Practical Knowledge

A Key To Living A More Functional Life

This book was professionally typeset on Reedsy.
Find out more at reedsy.com

Contents

1

Introduction

Have you ever pondered, or wondered about how the learning process works for human beings? What are the differences between book knowledge and practical knowledge?What are the similarities about the two as well? How do these two forms of accumulating knowledge play into the broader concept of gaining knowledge about a given thing, or in general about all things for that matter. This is something that I have thought about quite a bit lately. So, I am writing this book with the intention of analyzing these questions among others.

Is practical knowledge more important than book knowledge?Do one of these forms of knowledge trump, or prove to be superior to the other, and if so, is that always the case?Or is it dependent on the field or topic one is studying, or trying to master, or obtain a large amount of knowledge about?

I'm sure most of you, if not all of you reading this, have heard of "The Rule of ten." It generally states that it takes 10,000 hours of practice, or repetitions at something to truly master it to where it becomes second nature or takes very little conscious thought (if any) to perform that task. The specific number of 10,000 is not exact or even important in relation to this concept. What does matter is the general idea that it takes a long amount of time or repetitions for the brain to adapt to something to the point at which it takes very little conscious thought to carry out the task or understand the concept. This is a widely known adage in many fields, particularly sports performance, which happens to be a particular area of interest to me. I happen to work as a fitness/performance trainer and health and wellness practitioner.

In many instances, athletes need physical application of concepts to truly perform well, no matter how much classroom or study/meeting time is spent going over video from a previous competition or practice. This would tend to have one lean towards the idea that practical application is of greater importance than non-practical study.If that is the case why do athletes still have film/video sessions in the classroom, which they have basically been doing since video/film recording has been invented and applied to sports/athletics? Also, would this same importance or level or hierarchy apply to all things, not just sports?

I wish to explore these types of questions in this book. I am not advocating for practical knowledge to be necessarily more or less important than book knowledge or vice versa.I am an avid reader and have close to 100 books in my possession (most of which I have not read just yet).Reading is one of my favorite things to do. I do, however, think that reading (acquiring book knowledge, as important as it is, is just part of the process of learning and mastering any subject). I feel to truly grasp any concept or subject, actually getting your hands dirty and getting down to the 'nuts and bolts' of something, so to speak, is a necessary part of the learning process.

I will be going over some of my personal and professional experiences that led me to this conclusion of how vitally important practical application is in not only my career, but multiple different disciplines/fields, as well as just in life in a general sense. I truly believe embracing the idea of jumping into a situation in life can teach one quite a bit. I say that, with the caveat that learning some sort of book or study knowledge beforehand will only help prepare anyone for a less bumpy road traveled through that practical experience.

Through my 46 years on this earth, I have made countless bad decisions, as well as some good ones. In that time, I learned from experiencing things in life as much or more as just being told something or reading about that thing. Practical experience does not necessarily mean experiencing something yourself, you can also witness someone close to you experiencing it in a way where you were able to have a personal account of something, rather than hearing about it in a secondhand way. I hope the following chapters and content of this book helps everyone navigate their life in a more functional and productive way. I would like for the youth especially to not have to waste time as I did to figure out some of the hard truths in life and be able to figure out their path in life much sooner than I did.

It took me some time to figure out what I truly wanted to do in life. I went back to school originally in my early 30s rather than right after high school, and just recently finished my education/schooling by earning a master's degree at around age 44-45. It took this long due to having to take breaks from class at various times for various reasons, so I am constantly feeling like I am chasing time and trying to squeeze as much knowledge and accomplishments in as I can into the amount of time I have left on this earth. It has been said that "time is our greatest resource, because it is limited."In my case, since I feel like I had a late start in certain aspects in life, pursuing my higher education and figuring out a career path being one of those things, that is maybe more of the case for myself than the average person. Getting more practical experience earlier in life could hopefully help others avoid the stress of always feeling like there is not enough time to accomplish what is needed and wanted to be accomplished.

I hope you enjoy this short book on this subject. I wish for this to be a thought-provoking journey of practical setting experiences and book knowledge obtained by the reader.The hope is that this increases your desire to learn through reading as well as understand that there is no substitute for real world experience in pretty much any endeavor or subject you would wish to explore in your life. Whether you would want to learn about a specific field of study or career choice, or you just want to be a much more well-rounded individual in your journey of living, take the words in this work of literature as something you look to consider while navigating this crazy and interesting existence as human beings that we call life! Since I have discovered this revelation in my life about how important having a well-rounded learning experience in life is, I feel the need, obligation, and duty to pass this information on to as many people that will listen, or in this case read.As is the theme of this book, I would like for every reader to not only read this information but attempt to apply it practically to any and every facet of your lives. I truly hope you enjoy what is to follow and thank you for taking the time to go on this journey along with me.

2

My Personal Experience

I spent a large part of my life aimlessly just existing and not having much direction, or much of a future to speak of. That is until I experienced a life-changing pivotal moment when my father unexpectedly died. Unfortunately, my dad let himself go and did not take care of himself after falling into a state of depression that I had to witness firsthand. We were close when I was young but went through a phase where that was not the case when things got a little rough at home during my middle school and high school years, that also carried into my young adulthood in my early 20's. My father was laid off from work a few times and that was hard for him to process and get through, and he tended to be frustrated often. We all handle things differently, and he was not the easiest to deal with at that period. I did not understand what he was experiencing at that age. Only after I got older, did I understand better how hard something like that is to deal with, particularly if you have kids.

Some time passed and our rocky relationship as father and son changed, and we became close once again around the time I was reaching my mid-twenties. Unfortunately, this was short lived, because within a year or two of our relationship becoming strengthened again, he wound up passing away. He reached a state of depression which led him to begin drinking regularly, not being very active, smoking a pack of cigarettes per day, not eating the healthiest things, and on top of that he stopped taking his blood pressure medication. This all added up to a laundry list of risk factors that negatively affected his health, and he eventually wound up experiencing a heart aneurysm while driving.

One day I received a knock on my door from the local detectives, who then asked if they could come in, and when I asked what this was about, they said it was about my father and that I better sit down. Obviously, when you hear something like that, the news you are about to hear, or what you are about to find out is not usually good.They proceeded to show me pictures from the scene of my father's car accident where he was clearly deceased with his hair a mess and blood on his face from the impact of the collision with the light pole that he hit after losing consciousness from the blood vessel in his heart that burst from the aneurysm. You could imagine that this is a very jarring way to learn about a loved one's death, although hearing about something like that is never easy regardless.

After overcoming the obvious initial shock and going through the grieving process, I looked at this as an extreme wake-up call for what was going on in my life at the time. I did not have quite the same amount of risk factors as my father at the time (given that I was around 25 years old), but I was a heavy cigarette smoker and drank heavily on the weekend. My food choices were poor as well, and other than having a manual labor-based job, I was not very active. I did not exercise regularly and had a ton of stress in my life. So, I decided to join a gym that a close childhood friend of mine was a member of and began working out with him. We are still close to this day. I am proud to say we have a friendship now of over 35 years. That is a rare thing!

I became passionate about exercise to the point that I hired a fitness trainer and during my training with this trainer I became intrigued with the science related to health and fitness of the human body. I decided that I wanted to do this for a living and become a trainer so I could help people to avoid what happened to my father. It became an entire lifestyle change for me. I was able to quit smoking and the other bad habits and prioritize my health and fitness. I wanted to apply this newfound lifestyle change to helping others as to avoid the road my father went down and turn his tragic death into something positive for people.

Soon I purchased the materials from an online website of a fitness training certification company and began studying what I needed to know to pass a certification exam to become a certified personal/fitness trainer. From there I started training clients at a gym where I was hired, but I found out that it was much more challenging to do this in a practical way than it was in theory. I was lucky enough to become friends with someone right away (who is one of my closest friends to this day and is more like a brother to me than a friend) who was a much more experienced trainer than I was at the time, and he was able to guide me to a great degree through my development as a fitness trainer. He was able to eventually get me into a private fitness training studio that he trained his clients at where I would be around more experienced trainers like him who were able to apply training concepts that addressed their clients' needs more specifically. I was able to pick up some things to add to my repertoire as a trainer because of being in that environment, but I still was not satisfied with my understanding of everything. I felt my anatomy knowledge could be greatly improved along with other aspects of my knowledge base. So, I decided to go the academic degree route with my educational path rather than trying to study these things on my own in order to improve at my craft. I felt as though I could benefit from the structure and guidance from professors in a school setting.

I began my undergraduate program in a school nearby and completed an associate degree in science with a major in Fitness and Exercise after a little less than two years. After being away from class for some time (longer than I wanted), I then went to another institution that I felt had a better program for completing my bachelor's degree in Exercise Science. This long hiatus from being in school was because I chose to take advantage of a study abroad opportunity in Florence, Italy near the end of my first stint at my first school just before acquiring my associate degree. I knew there was a chance that this study abroad program would set back my timeline for completing my schooling due to the cost and the process of applying for a different program at a different school as my next step, but I felt the experience that I would have gone through would be worth the risk. Even though I did not have the funds to pay for the trip, the school allowed me to put the cost on my student account and I would likely have to repay the amount before obtaining my transcripts to apply for the other program at the different institution. However, I still thought it was worth the gamble to experience a once in a lifetime opportunity in a sought-after destination such as Italy for a 6-week period.It turned out I was right. To this day I still have no regrets about the decision I made, and I would do it again the very same way if I had the opportunity again, even though it put me out of class for a little over 5 years in order to repay the debt and apply for the program at the other school. This experience was pivotal in my learning process for my studies, but also gave me exposure to another culture and way of life.

I learned some very useful things at my next stop at the second educational institution I attended during my bachelor's degree studies, but I still was not satisfied with where my knowledge was at. When I started this program, I had no intention of pursuing a Master's degree, but sometimes things change. While researching for an assignment near the end of my time pursuing the bachelor degree, I came across information about a program at a third school offering a Master's Degree program in Kinesiology (Human Movement Science), with a concentration in 'Human Performance.' The cost seemed reasonable, and they offered a timeline where even if I took one class at a time, I could complete the program in 2 years or less. I had my heart set on becoming a trainer of athletes at this moment in time because of my love of sports and athletics and due to being a little turned off by some of the practices and ways the general fitness industry seemed to be heading in.I felt like there were too many gimmicks and marketing ploys that ignored an often-used phrase 'Best Practices' in the fitness realm, especially after the arrival of social media with so many fitness influencers. I started seeing things that looked like they were terrible for people to be doing, that most people without a background in the field would not know any better about.I really did not like this at all.I thought that the strength and conditioning industry held higher standards and truly had the 'best practices' in mind being that they trained athletes exclusively. It turns out that I would later find that not to be the case…. I will get into more detail about that later in this book.

At this point I began my master's degree program. I was thrilled to be accepted in the program and even more thrilled to start really learning advanced performance exercise theory.The school was in another state and the program was fully on-line. This was during the time of the Covid-19 Pandemic, so this was something that made sense at the time with remote learning being so prevalent. A lot of the classes involved research article assessment and discussion board interaction with the rest of the class and writing papers on things related to performance training with athletes. There were final projects in these classes that at times would get more practical but being that it was an online program there were limitations to how practical things would be allowed to get. I did very well in the program and managed a 4.0 GPA for the duration of my time there. I learned quite a bit from the class material, especially things that related to research articles, and everything that is involved with how data is gathered, variables, and how the data was tested, as well as the population or groups used as sample sizes to do the studies.

I found it particularly interesting that one of the first things I learned in a research class was how to spot biases and other things that could rig studies. I was amazed to find out how many studies were manipulated to prove a per-existing conclusion or a hypothesis a researcher already had come to before starting the study and what they would do to appear to prove that belief/hypothesis.There were some assignments in classes where I would post videos of assessments on clients/volunteers but for the most part it was research and book learning. I found this helpful, and I felt I learned a good amount but something was still missing.

3

What I Found Changed My Entire Outlook

A short time before I finished my undergraduate program prior to starting my Graduate degree program COVID-19 happened, which I do not have to tell anyone was a big adjustment. I was in an area where the lock downs and COVID Protocols were a bit more on the strict side, so I found myself with a lot of time on my hands, mostly being stuck in the house. This led to a good amount of time on-line. Prior to this I stumbled upon a fitness company named 'Functional Patterns.' For some time prior to 2020 when COVID happened, I would see informational posts about their training concepts and methodologies that really intrigued me, but I did not have the time to really dive into the physical training of this method by experiencing it on myself or any clients before COVID happened. Now that I had the time, I really dove into the material and it changed my whole outlook on training the human body, and later it would change my outlook on life and through that I was able to see a more clear route to living a more functional life because of it. I was able to experience this method first hand and it was far different from conventional training practices. It concentrated on really correcting the foundation of the human body's structure from a postural standpoint (something that is just usually briefly covered in fitness training certification manuals, but not really prioritized by most trainers in the field). This is accomplished through tensioning the body correctly from a static posture position, then progressed through corrective exercise from different positions/settings, before going into actual dynamic movement/strength exercises.

Going through the program and feeling the changes happen to my body directly (decreases in joint discomfort and pain, structural improvements in my static and dynamic posture, and overall being able to move a lot better and more functionally) really had an impact on me beyond all the years of school I just spent a decade on and off going through. I do not regret my path in school or feel as if it was worthless by any means but this program from the people at this Functional Patterns company really made me understand everything I was learning a lot better and I also learned new things that are not taught in conventional exercise science programs for degree programs or certifications.

This company was able to do things beyond the rest of the industry because they were not afraid to look at things with a critical eye and physically and practically test things out in real life. They focused heavily on the principles of physics and bio mechanics based on how humans have seemed to evolve to move and function in order to survive over the years and become more functional over that time. A lot of the things the fitness/rehab/performance industry does are based on things that have just been passed down over many decades and nobody tends to question the validity of these things. If influential people in the industry who have a lot of initials after their name and are prominent figures high up on the hierarchy of certain fitness certification company's board of decision-makers on what principles to teach vouch for something, or pass on certain practices, the information is not really questioned.

Most people see the credentials and just accept everything as 'gospel.' A lot of this information is gathered through studies that are, to be very blunt, flawed. In general this industry has been very resistant historically to making dramatic changes to their education protocols. Traditional practices that were in practice decades ago have not changed all that much, particularly in the strength and conditioning community that is responsible for certifying trainers of athletes. Little small details may be added, but they are very reluctant to scrap certain training concepts (such as conventional weight lifting), in my opinion due to an emotional attachment to certain things and reluctance to admit certain things are not as good for the body as originally thought. People seem to not want to step on toes or offend people. Well, life can be harsh at times. If we really want to use 'best practices', that means some things may need to be changed drastically.

Studies are great but they are looking at things in isolation to see one or two specific variables and how they work. Well, real life has countless things interacting with select variables. So, the takeaway I want people to experience from this should be that things cannot be as controlled in practical application out in the real world as they are in research studies in laboratory settings. Also, most studies, at least in the fitness sciences realm are not longitudinal studies that gauge chronic (longer term) affects as much as acute ones. These people at "Functional Patterns" were not afraid to look at current 'best practices' with a critical eye and ask some probing questions. They did their own research and tested their own protocols over thousands of hours, and eventually were able to get very tangible anecdotal results with before and after images and testimonials of people's lives improving at a very high rate that were not achieved with virtually every other method of training. A true test of the validity of a training method in terms of fitness, performance, and health would be if the adaptations/effects of the training stimuli garners great results for the average person, rather than elite athletes who could be achieving their results, at least partly based on being blessed with great genetics. People who are not naturally athletic who are able to make significant athletic gains is more of a valid result

My eyes were also opened to other things through book recommendations from the founder of this company, Naudi Aguilar. I read a book called "Tyranny of Words" by Stuart Chase that was one of the recommendations. This book would have a big impact on my perception of how things are presented versus how they actually are in reality. I would recommend anyone reading this to read that book also. One of the many quotes that stuck with me from this book was, *"People are not 'dumb' because they lack mental equipment; they are 'dumb' because they lack an adequate method for the use of that equipment. (Chase, SC, 1938, The Tyranny of Words, P. 28)"* This book is somewhat dated but makes some great points. Sometimes going back and looking at things from the past can be just as useful as more present work.

Another one of the book recommendations that I followed through on was a book called, 'The Mind in Making.' One of the quotes from this book that stayed with me is, *"Partisanship is our greatest curse. We too readily assume that everything has two sides and that it is our duty to be on one or the other* (Robinson, JHR, 1921, *The Mind in Making, P. 225)."* This pointed to the fact that some things aren't as much a matter of opinion as much as being factual/provable or not. Our feelings and emotions can get ourselves in trouble at times. We must be able to put aside our hubris/ego if we are truly to find the truth, or what is more likely to be factual and the best course of action. Pointing out the truth is not necessarily an attack on an opinion that someone has from a different standpoint. Experiencing things in a practical way can oftentimes point us in the right direction to finding the truth. This is why I advocate for acquiring practical knowledge as much as I do.

If you noticed, I referenced some books just now. That goes back to my stance from the very beginning that book knowledge is helpful, I am not here to bash acquiring book knowledge, but it is part of the answer, not the end-all/be-all. I love learning, and books are a huge part of that learning process for me, but that process does not stop there if one truly wants to have vast knowledge about any given topic or subject. I believe that context and setting is truly an important aspect of real understanding of anything. These are a lot of the things Naudi and the people at 'Functional Patterns' stress to their audience. I would suggest that everyone reading this also look at a book called "The Power of Posture" written by Naudi Aguilar, and check out their company website WWW.FunctinalPatterns.com. They really do some incredible things for people.

One thing I have come to realize, particularly more recently, is that many things are not the way they seem. To really realize this you need to experience life. There is a saying that goes something like, "experience is the best teacher." This is a very true and poignant concept. It is easy to draw a conclusion about something that could be flawed by observing it from a distance. We truly find out about things from experiencing them in a more up-close and personal way.

4

Nuance To The Discussion

Like anything though, there is nuance to this discussion about book knowledge versus practical knowledge. For instance, if you are studying or reading about a different culture (Asian culture for instance is something that I am interested in and I have read various books on Japanese culture specifically), you may not be able to experience Japanese culture in person if you live on the other side of the globe like I do. So, for many people in a case like that, practical knowledge in person may not be an option, so book knowledge and reading about that may be your only recourse to gain any understanding, or at least that would be the case for a significant amount of time until you had the means to make a trip like that. So again, I am not advocating for book knowledge being unnecessary or completely useless, just that practical knowledge/experience gets you a different perspective that is necessary if you wish to have a certain level of mastery of something or more advanced understanding of it.

Look at various different fields of science for instance.It was actually said that Sir Isac Newton discovered the existence of gravity by simply watching an apple fall from a tree. So that revelation was come to by interacting physically with his natural environment (practical, real world interaction). Of course after that point many books were written that explain the concept of gravity. There is a specific case of the importance of both types of knowledge being used. In this case, it would be practical knowledge (discovery) first, followed by that discovery being put into books for educational purposes through book knowledge. Logically then, many science students/professors then conducted experiments to experience this phenomena in a practical, real world setting.

Physics in particular seems to be a field that is greatly based on practical knowledge/application since it is the study of things that happen in our natural world in terms of internal and external forces and how they are manipulated. Although you could argue that learning about these things is greatly dependent on book learning as well. Biology on the other hand, may be perceived as something that is learned more from books aside from microscope observation. It is not something the naked eye can witness, which makes it harder to observe in the natural world. So book knowledge in this example could be more greatly utilized as a learning tool for Biological sciences. I would say that most things that occur in our natural world are understood at a higher level once practical knowledge is obtained through physical interaction of some sort though.

The nuance aspect to this particular part of the discussion is based on the premise that there is a natural ebb and flow to things. In other words, not everything is an 'open and shut case' so to speak. Some things in life/study rely more heavily on book knowledge due to a lack of accessibility to certain things or places physically. This is understandable, but it should also be acknowledged that most things do not fall into this category so this should be viewed as the exception rather than the rule. Most things we learn about can be experienced in a more practical, real life physical type of way. This is definitely true when talking about our career choices. We learn about our field greatly from book knowledge, but many times we find that no matter how much we study about the job we are about to embark on beforehand, we do not have the great understanding of it that we eventually gain during the physical experience in that field.

5

Natural/Universal Law

The Universal Laws are also known as the 'physical laws that govern all things in our reality.' Some of these laws are; "the law of polarity, the law of vibration, attraction, action, cause and effect, rhythm, etc. These are things that let us know that our physical interaction with things not only tell us quite a bit about our reality, but actually put us on a higher plane of understanding when we are physically immersed in our environment. All of these concepts relate to physical interaction with our natural world. These tidbits of knowledge are said to have come from Ancient Egyptian and Greek cultures/societies. People who have studied the construction of the ancient pyramids in Egypt reference these natural or universal laws to explain how and why these structures were built. Seeing pictures of these things and analyzing the shapes are useful from book knowledge, but this is another instance where you could imagine physically being in the presence of these marvels of construction and architecture elicit a whole different level of understanding.

This is one of the main reasons why physical experiments

became a real thing in the field of scientific study. All those years ago when physical experiments were first implemented, the importance of practical application was acknowledged. So, it is my contention that different forms of learning are important to grasping various concepts. None however, may be more important than the form of practical learning/knowledge. We as human beings have 5 senses that make us aware of things in a very specific way. Those senses can be more fully activated through physical contact with other things. Our senses enhance the messages sent and received in the human nervous system. This enhances cognitive function within all of us.

Many different forms of learning encompass how we gather knowledge. It has been said that some of us are more auditory learners (learning by listening/hearing), some are more visual learners (learning from sight), reading/writing (notation and reading words), and others are more of what is referred to as kinesthetic learners (practical experiment based/witnessing). This is known as the "VARK" model of learning. This was designed by a man by the name of Neil Fleming in 1987, (Mozaffari, H.R., 2020, *The Relationship Between the VARK Learning Styles and Academic Achievement in Dental Studies).* There are some other models that suggest there are 7 different types of learning, but for the purposes of keeping this analysis more simplistic, we will look at the 4 type 'VARK Model.'

Behavioral scientists who study how human beings learn have come to the conclusion that most of us learn in all of these ways, or with some combination of all of these things, but some rely more on a single form than others. I would argue that most of us probably absorb the most understanding from the latter 'kinesthetic learning." Humans by nature are very much communal or social creatures. We crave or require physical interaction with other things or beings to a large degree. When we physically interact with our surroundings we experience a sense of feel/touch (sensing if an object is smooth or rough). We can hear and smell things that allow us to draw conclusions.We also get to see things in person.When we are reading about things we often need to use our imagination to imagine things that we are reading from a descriptive standpoint. Even if we are provided illustrations or actual pictures in a book or article we read, it is not quite seen with the same amount of accuracy as physically being present to the subject in person. Not being present allows for us to imagine exaggerated things at times, or even imaginary things that are not present due to us not technically being present in the situation. Physically witnessing things in person greatly reduces or can even eliminate this possibility of 'letting our imagination run away with us' so to speak.

6

Mathematical Knowledge

Mathematics might be one of the few fields where I feel like book knowledge is possibly as impactful as practical knowledge. Although you could argue that solving math problems with a pencil and paper could be considered a form of practical application. You would not have to physically have items in your hand that you need to subtract or add for it to necessarily be considered practical application in this case, so there is also an argument for mathematics being a discipline or subject where practical knowledge is also a key to unlocking more significant understanding. If you think back to math class in school, you tended to understand the concepts/theorems in a more in depth way when working out the math problems at the end of the chapters after the lecture section of the books.

Math and science probably are the two subjects that have the biggest effect on a person's life. Without those two subjects being understood well, there is a good chance that your functionality in life will suffer. I was always a student who did better in English rather than Science and Math. I was a fair student in science as a kid, but math was always a subject I struggled in more than any other subject. I think like a lot of kids, I had some anxiety about mathematics due to seeing all the large figures and lengthy formulas/equations. We can let ourselves get into a panicked state easier than we should at times.

Math is actually a subject that I have gotten into much more lately after having an epiphany about its true importance. This importance is not just for the ability to do things like, "balance your checkbook", yes, I know I dated myself with that phrase, but also the importance of numbers in a more elevated way. I was made aware that certain numbers are associated with things in our natural world and have deeper meanings through some more recent study. Recently I have read two books ('PhiloMath and Polymath) that dove into the connection with certain forms of sciences and mathematics or number theory, specifically how geometric figures that shape our world are tied to numbers. This knowledge was presented to me in a book knowledge format, but in that book there were practical examples from our real world that connections were made to.

Numbers 3, 6, 9, and 7 all have a significant amount of importance according to the author of these books. This is also not the first time I have heard of such things. Individual numbers and series of numbers can have very deep meanings and implications on our reality. This is a widely accepted concept. The book "PhiloMath" has a section of particular importance that talks about the divine nature of the number 7, *"It is the number of prime colors, as well as the 7 primary planets, the 7 heavens, 7 virtues, 7 internal organs, 7 chakras (energy points). Even the Great Pyramid of Giza, some argue, is built upon the geometric representation of number 7, the heptagon (Grant, R.E.G., 2021, Philomath, P. 86)."* You can see how these are mostly practical things that can be seen or observed in a physical way if one had the means to do so…aside from 'chakras.' Being able to,if you so wished, to physically witness all these examples of the number 7 is a great example of book knowledge being a precursor to practical knowledge, and how both forms of learning or acquiring knowledge are vitally important to the process of learning. I hope whoever is reading this can see how showing how both book learning and practical experience can be unified and integrated to give someone a better overall understanding of a given subject or area of study.

7

Our Over-Reliance On Research Study Data

One thing I have noticed is that we are becoming dangerously close to developing a dysfunctional over-reliance on data from research studies. It is as if we feel the need to see a data set from research studies about the littlest and most obvious things. When I witness people having discussions, debates, or conversations about various subjects I see evidence of this. Do not get me wrong, I work in a field heavily reliant on data from studies so I see that there is some importance to it, but if someone told me the sky was blue, I would not say, "can you show me the results of the study that prove that?" I would simply look up towards the sky on a clear sunny day and clearly see the sky have a blue color if it were not cloudy on that particular day.

I am not quite sure exactly when anecdotal evidence became such a taboo thing, but somewhere along the line that has seemed to happen to some degree. It must have gotten lost somewhere that anecdotes were some of the first data forms of early scientific principles that we hold to have great importance today. I mentioned earlier that Newton's theory of gravity was discovered by chance to some extent when Sir Isacc witnessed an apple fall from a tree it was growing on. This is an example of anecdotal evidence. This also spawned the calculation of 9.8 meters per second squared that we associate with the speed of gravity or gravitational pull here on earth. I would like to add that this was obviously proven or found to be the case by practical experimentation. This would be the only way to positively come to the conclusion that this is the speed at which objects fall or are pulled down toward the ground on this planet. This is yet another example of the importance of practical forms of knowledge or learning.

The simple act of observation in real life practical situations can tell us a lot if we know what we are looking for and have a trained eye to look for it. We can learn quite a bit from what we witness in a natural setting. This is why I borrow a phrase from the people at Functional Patterns, "Become a doer more than a thinker." Doing comes from first thinking and having a thought about something in mind, but you cannot prove your theories or have those theories evolve and get molded into a more elevated place without testing out your thoughts in real life situations. Laboratory research experiments give you some results, but they simply cannot account for enough real world variables in such a controlled environment. There are a lot of different things at play in reality. Too many combinations of stimuli exist in the practical world to simulate that in a laboratory setting.

This over-reliance on data from controlled laboratory research is particularly an epidemic in my field of fitness/performance/rehab/health and wellness. Part of the problem is that there are not many, if any at all 'Longitudinal Studies', or studies that show data over long periods of time. It is important especially in this field to show more chronic adaptations than acute ones due to the long term effects related to injuries due to training stimulus over time. This is something that Naudi Aguilar (the founder of Functional Patterns) has mentioned countless times in interviews and is the subject of material for his company's website. This was also the topic of an episode of 'The Art of Move' podcast that discusses movement/biomechanical science. If anyone reading this has the time I would advise them to watch the podcast on YouTube. It is a really great discussion on human biomechanics that is applicable to most all walks of life or subjects as well.

After being exposed to this reality of our over-reliance on research data, I really looked at my industry, and life in general in a different way, and was able to really experience a paradigm shift (change in perception) to things in general. We must be careful not to rely on study data so much to where what is right in front of us is not even seen! I feel this extends beyond our perception of scientific studies. If we are not careful this effect could bleed into every facet of our lives and reality. One of the drawbacks to all of our modern conveniences of comfort and shelter, is that we no longer interact with our natural environment anywhere close to the degree that we used to many years ago. This limits our ability to experience things more physically in their natural state, which I believe is the intended way to experience life in general.

8

Conclusion

The biggest take away that I want the reader to have from this short work of mine is that although book reading/learning is still a vital tool in understanding anything, do not underestimate the effect of physical experience either. Life is something that we have to physically experience in every way possible in my humble opinion, if we are to acquire a significant amount of useful knowledge. A popular phrase is, "Knowledge is power.", and the way we absorb the best and largest possible amount of information is through actual experience. This is something that I feel we need to be particularly mindful of in the coming years, even more so than before due to a lot more automation being implemented with the added use of things like AI (Artificial Intelligence).

If we are not careful, being complacent and borderline lazy with how we choose to acquire information will hurt us in immeasurable ways as a society in the future. Do not get me wrong, I am not of the opinion that AI is a bad thing. In fact, there are tons of useful ways that this technology will be implemented to improve our quality of life, but we need to make sure we still have our humanity and are in control of our lives to some extent at least. A quote from another book comes to mind, *"There is but one remedy, one hope – a science and art of Human Engineering based upon a just conception of humanity as the time-binding class of life and conforming to the laws of nature including the laws of human nature* (Korkzybski, A., 1921, P. 118, *Manhood of Humanity).*" I believe that the part of this quote that mentions 'Human Nature' is a very important concept. One that we must hold on to.

The moment we lose our humanity, which is deeply rooted in physical interaction with others, as well as our natural environment (nature more specifically), that is when our ability to continue to lead as a species will suffer. We are witnessing a deterioration of thought that most of us do not even realize is actually happening to us. We are being told what and how to think in general more now than ever, rather than just given the information and being left to ourselves as how to best piece that information together, and come to our own conclusions. We have a lot of distractions to deal with currently that did not exist prior to these current times. We need to be brave enough to still look at things critically, and question anything and everything. It is not a crime nor is it an accurate indictment of being insensitive to want to have clarity or accurate information about any given subject, no matter how trivial it seems to be from any one person's perspective!

Part of the beauty of life is that we get to actually go out and physically experience it ourselves. Yes, there is a fear of making mistakes, but as the saying goes, "we live and we learn." In other words, do not be afraid to make mistakes. There has never been a person that ever lived who did not make some mistake at some point in time. Many times, some of our greatest moments of personal growth and breakthrough come from some of the biggest mistakes that we make. Never be afraid to fail or look 'stupid' in anyone's eyes. Without failure, you really cannot become a winner. Usually when one makes a mistake, they are wiser for it and the mistake is not made again, especially if it was a big mistake. That is most likely because you remember the way you felt when you made that mistake and you do not want to experience that feeling of disappointment or despair again. Keep reading as much as you can to learn new things and test out those new tidbits of information through practical application of those things you obtained through book/reading knowledge. You will be glad you did, if for no other reason than to confirm the information you received through the book knowledge you gained. There are very few greater feelings than having a belief, and then having that belief or understanding be confirmed in a very real and practical way!

9

Resources

Aguilar, N. (2014). *The power of posture.*

Chase, S. (1938). *The tyranny of words.* https://pure.mpg.de/p
ubman/item/item_2309475_4/component/file_2453209/Cha
se_1938_Tyranny.pdf

Grant, R. E. (n.d.). *Philomath: The Geometric Unification of
Science & Art Through Number.*

Korzybski, A. (2021). *Manhood of humanity.* http://lipn.univ-
paris13.fr/~duchamp/Books&more/Neurosciences/Korzybski/
%5bAlfred_Korzybski%5d_Manhood_of_Humanity(BookFi.o
rg).pdf

Mozaffari, H. R., Janatolmakan, M., Sharifi, R., Ghandinejad,
F., Andayeshgar, B., & Khatony, A. (2020). <p>The Relationship
Between the VARK Learning Styles and Academic Achievement
in Dental Students</p> *Advances in Medical Education and Practice, Volume 11,* 15−19. https://doi.org/10.2147/amep.s235002

Robinson, J. H. (2012). *The Mind in the Making.* BoD − Books
on Demand.

About the Author

Brian Gardner has a Master's Degree in Kinesiology (MSK) with a concentration in Human Performance. He works as a fitness facility manager and fitness/performance trainer and health and wellness practitioner. He is an avid reader and now writer. His interest/field of study ranges from everything to self-improvement to every form of scientific study (psychology, biology, quantum biology/circadian health, exercise physiology, human biomechanics/physics), as well as coaching and motivating others, and mathematical study, theology, and philosophy. He is dedicated to living a more functional life and coaching/training/leading others to do so. He has plans to someday soon own and operate his own training facility that develops human performance.